OUR SOLAR SYSTEM

Comets, Asteroids, and Meteoroids

BY DANA MEACHEN RAU

Content Adviser: Dr. Stanley P. Jones, Assistant Director, Washington, D.C., Operations, NASA Classroom of the Future

Science Adviser: Terrence E. Young Jr., M.Ed., M.L.S., Jefferson Parish (La.) Public Schools

Reading Adviser: Dr. Linda D. Labbo, Department of Reading Education, College of Education, The University of Georgia

COMPASS POINT BOOKS

MINNEAPOLIS, MINNESOTA

Compass Point Books
3109 West 50th Street, #115
Minneapolis, MN 55410

Visit Compass Point Books on the Internet at *www.compasspointbooks.com*
or e-mail your request to *custserv@compasspointbooks.com*

Editors: E. Russell Primm, Emily J. Dolbear, and Catherine Neitge
Photo Researchers: Svetlana Zhurkina and Marcie Spence
Photo Selector: Linda S. Koutris
Designer: The Design Lab
Illustrator: Graphicstock

Library of Congress Cataloging-in-Publication Data
Rau, Dana Meachen, 1971–
    Comets, asteroids, and meteoroids / by Dana Rau.
        p. cm.— (Our solar system)
Includes bibliographical references and index.
Contents: The beginning of the solar system—Comets: dirty balls of ice—Comets: a long journey—Asteroids: minor planets—Asteroids: where they orbit—Meteoroids and meteors: rock from space to earth—Meteorites: hitting earth's surface—Studying the solar system—Our solar system.
    ISBN 0-7565-0437-6 (hardcover)
    1. Comets—Juvenile literature. 2. Asteroids—Juvenile literature. 3. Meteors—Juvenile literature. [1. Comets. 2. Asteroids. 3. Meteors.] I. Title.
    QB721.5 .R38 2003
    523.6—dc21   2002009937

Table of Contents

NOTE: In this book, words that are defined in the glossary are in **bold** the first time they appear in the text.

The Beginning of the Solar System

The solar system is a very large place. Nine planets, including Earth, travel around the Sun, or revolve, in paths called orbits. Many of the planets also have moons. Other objects are a part of our solar system, too. They are called **comets**, **asteroids**, and **meteoroids**.

Astronomers have a theory, or idea, about where comets, asteroids, and meteoroids came from. They believe that these objects are leftover pieces of rock and ice from when the solar system formed.

Astronomers think that the solar system began more than 4 billion years ago. It began as a huge cloud of dust and gas. This cloud started spinning. Most of the dust and gas moved to the center of the cloud and became very hot. This formed the Sun. As the cloud kept spinning, the rest of the dust and gas spread out from the center. Pieces of this material started sticking together. The clumps grew bigger and bigger. They became the planets and their moons.

◀ *The spacecraft* Pioneer II *flying past Saturn, one of the nine planets that orbit the Sun*

Some of the material never became planets or moons. The smaller clumps became comets, asteroids, or meteoroids.

The small white particle in the center of this circle of gas might become a comet, asteroid, or meteoroid. ▶

This brightly colored cloud of gas may resemble what our solar system looked like when it began billions of years ago. ▼

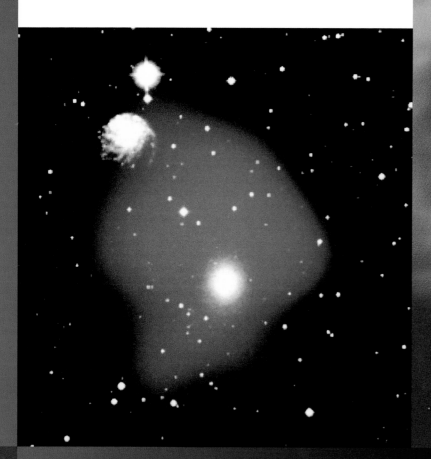

Comets: Dirty Balls of Ice

⭐ From Earth, a comet looks like a bright ball followed by a tail of light. Comets were called "hairy stars" by people who lived in Greece long ago. Thousands of comets have been discovered.

Comets are usually about 1 mile (1.6 kilometers) to 10 miles (16 km) across. The main part of the comet is called the nucleus. It is made of solid ice with bits of rock mixed in. Because of this, some people call comets "dirty snowballs." A cloud

◀ *Ancient Greeks called comets "hairy stars."*

of gas, called a coma, surrounds the nucleus. The coma is surrounded by another large area of gas. This gas is called hydrogen.

The brightest part of a comet is its tail. When the comet travels close to the Sun, some of its ice becomes gas. The wind from the Sun blows the gas to form a tail. Tails can be very long. The tail of one comet stretched 300 million miles (483 million km).

Many people who lived long ago wrote about seeing comets in the sky. In the

When comets travel close to the Sun, ▸
they may form long tails.

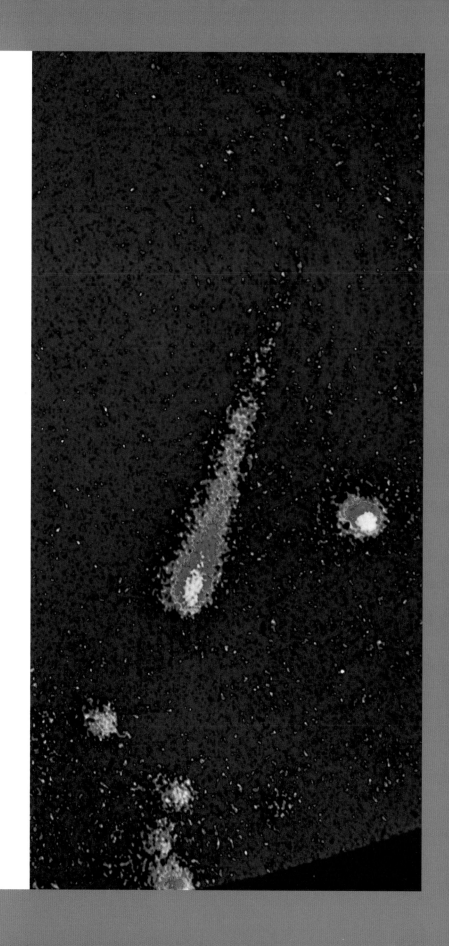

1700s, a man named Edmond Halley studied these writings. Halley noticed that for nearly two thousand years, people had seen a comet about every seventy-six years. He believed everyone was writing about the same comet. He thought that comets appear in the sky, go away for a while, and then come back again.

Halley needed to prove his idea. In 1705, he said that the comet would come back again in 1758. He was right! The comet was named after him. It is now called Halley's Comet.

How do comets come

◄ *Eighteenth-century astronomer Edmond Halley (1656–1742)*

back? A comet travels in an orbit around the Sun, much like the planets do. However, comets' orbits are different from planets' orbits. Comets travel in more elliptical, or oval-shaped, orbits. Sometimes their orbits take them very close to the Sun. Other times they are very far away from the Sun. These comets are too far away to be seen from Earth.

Like other comets and the nine planets, ▶
Halley's Comet orbits the Sun.

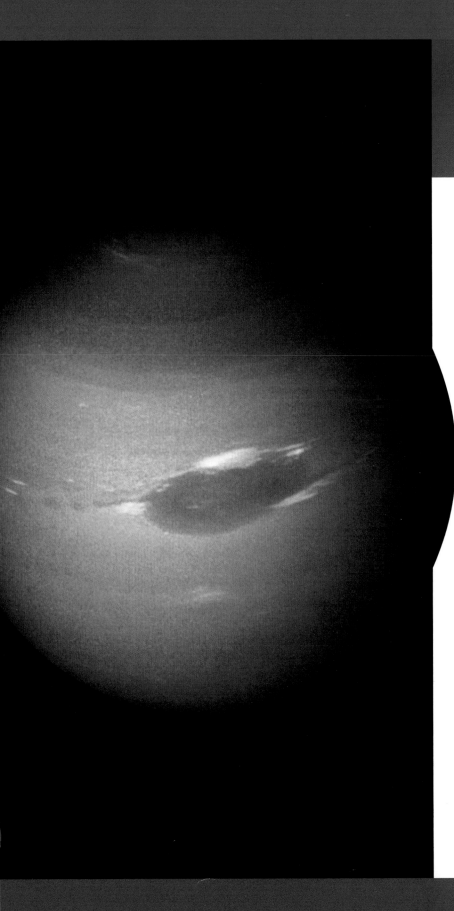

Comets:
A Long Journey

There are two kinds of comets. Short-period comets take less than two hundred years to complete their orbits around the Sun. Long-period comets may take as long as 30 million years to orbit the Sun.

The length of time it takes a comet to orbit depends on where it comes from. Astronomers believe that short-period comets come from the Kuiper Belt. It is an area beyond the orbit of Neptune. The Kuiper Belt is filled with

◀ *The Kuiper Belt lies beyond the orbit of Neptune, pictured here.*

many small, icy objects. These objects orbit the Sun together. Sometimes one of the objects leaves the belt and travels in its own orbit around the Sun. Then it becomes a comet.

Long-period comets come from the Oort Cloud, an area much farther away. The Oort Cloud is made up of as many as a trillion comets. It surrounds the entire solar system. The Oort Cloud is too far away to see. However, astronomers believe it is there. If a comet breaks free, it starts on its own orbit toward the

The Oort Cloud surrounds the ▼
solar system.

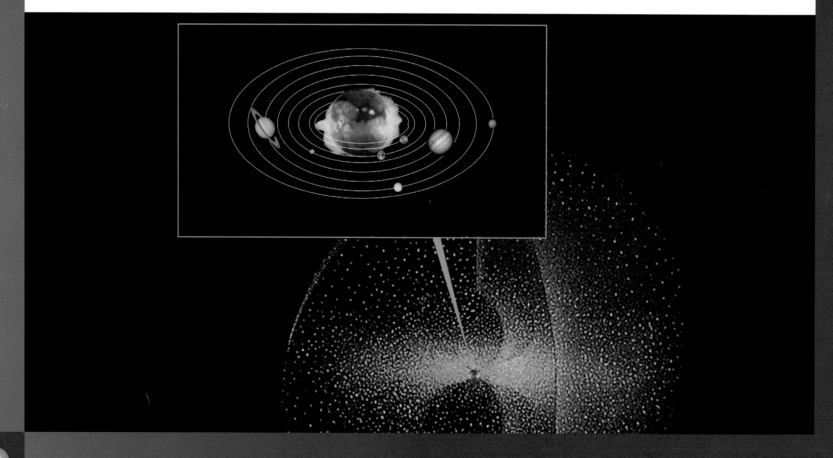

Sun. Because the Oort Cloud is so far away, the orbit is very long. It may take the comet thousands of years to make one trip.

Comets do not last forever. They lose a little ice and gas each time they pass the Sun. Over time, they become loose rocks that fall apart.

Some comets, called sun-grazers, get too close to the Sun and crash into it. Other comets crash into planets. In 1994, scientists watched as Comet Shoemaker-Levy 9 crashed into Jupiter.

▲ *A blue comet that has broken free from the Oort Cloud continues an orbit toward the Sun.*

◀ *Comet Shoemaker-Levy 9*

Asteroids: Minor Planets

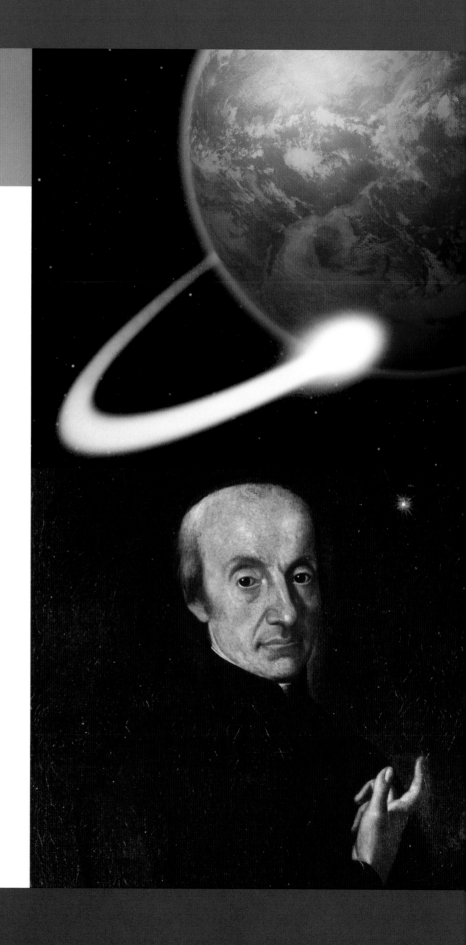

⭐ Like comets, asteroids are small objects that orbit the Sun. Unlike icy comets, however, asteroids are made of rock and metal. Also, asteroids do not orbit from as far away as comets. Their orbits are within the orbits of the planets.

Asteroids were first seen by the astronomer Giuseppe Piazzi in 1801. Astronomers had believed there might be

Unlike the icy comet shown here, asteroids are made of rock and metal. ▲

Giuseppe Piazzi thought he had ▶ *discovered another planet in 1801, but he actually had viewed a very large asteroid.*

another planet between the orbits of Mars and Jupiter. When Piazzi (1746–1826) looked through his telescope he saw a large object that he thought might be a planet. He called it Ceres.

It was much smaller than other planets. In the next few years, many more small objects were discovered. They were orbiting the Sun between the orbits of Mars and Jupiter.

These objects became known as minor planets. In 1802, the astronomer Sir William Herschel (1738–1822) named these objects asteroids. *Asteroid* means "starlike" in Greek.

Ceres is the largest asteroid. It is 580 miles (933 km) across. Most asteroids range from being less than 1 mile (about 1 km) to about 120 miles (193 km) across.

▲ *The largest asteroid in our solar system is Ceres.*

Asteroids: Where They Orbit

⁂ The area between the orbits of Mars and Jupiter holds millions of asteroids. It is called the Main Belt. There are also hundreds of asteroids called the Trojan asteroids. They travel with Jupiter in its orbit. Some travel ahead of the planet. Some travel behind it. There are also some asteroids that orbit between the planets Saturn and Neptune. These asteroids are called Centaurs.

Scientists know of about 250 asteroids that orbit much

The Main Belt of asteroids lies between ▶
the orbits of Mars and Jupiter.

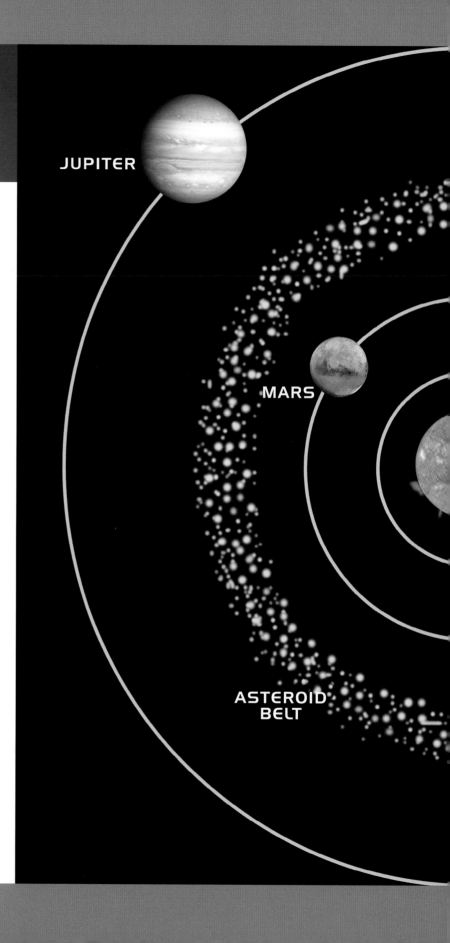

JUPITER

MARS

ASTEROID BELT

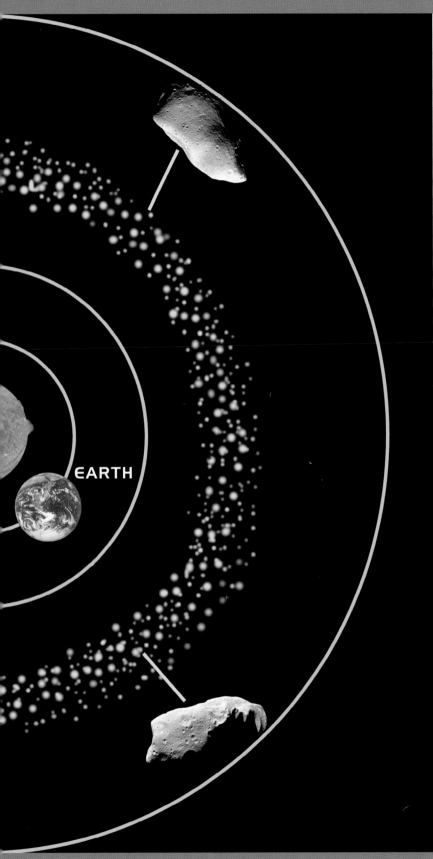

EARTH

closer to the Sun. These are called near-Earth asteroids (or NEAs). The orbits of the NEAs are close to Earth's orbit around the Sun. The orbits of some NEAs even cross Earth's orbit. It is possible that one

▼ *Ubehebe Crater in Death Valley, California, is evidence of asteroids hitting Earth.*

of these NEAs could hit Earth. Scientists know that asteroids have hit our planet many times in the past. The proof is the **craters**, or holes, they left behind.

Sometimes an asteroid that leaves its orbit may be "captured" by the **gravitational pull** of a planet. Mars's two moons, Phobos and Deimos, are believed to be captured asteroids. Some outer moons of Jupiter, Saturn, Uranus, and Neptune also may have once been asteroids.

Mars's moons Phobos (top) and Deimos (bottom) may have been asteroids. ▶

Meteoroids and Meteors: Rock from Space to Earth

✦ There are a lot of other pieces of dust and rock orbiting through space. If the piece is too small to be called a comet or an asteroid, it is called a meteoroid. Meteoroids may be pieces a comet left behind. They may be pieces from an asteroid. As Earth orbits the Sun, it hits these meteoroids.

Earth's surface is protected from these meteoroids by its atmosphere. An atmosphere is

▲ Some meteoroids form from pieces of asteroids, such as this one.

◀ A meteoroid approaches Earth, but it will probably burn up in the atmosphere before it gets close to the ground.

the mixture of gases that surrounds a planet. When meteoroids hit Earth, most of them burn up in the atmosphere. They never reach the ground. Astronomers think that each day Earth is hit with about 1,000 to 10,000 tons of rock from space.

When meteoroids hit the atmosphere and burn up, they are called **meteors**, or shooting stars. They look like streaks of light in the sky. Meteors can best be seen after midnight and right before the Sun comes up. They are more common in the fall and winter.

A meteor shower occurs when a lot of meteors can be seen at once. An especially large number of shooting stars is called a meteor storm. Some meteors are very bright and make a sound. They are called fireballs.

An illustration of a fireball from a popular ▼
astronomy book published in Paris in 1870

Meteorites: Hitting Earth's Surface

Not all meteors burn up in the atmosphere. Large ones may make it through the atmosphere and fall to the ground. They are called meteorites. Most of them fall into the ocean.

About 120 craters from meteorites have been found on land. The largest is the Barringer Crater in Arizona. Astronomers think it was formed about 50,000 years ago. It is about 3,937 feet

◄ Meteorites can create large craters, such as the Barringer Crater, when they hit Earth.

(1,200 meters) wide and 656 feet (200 m) deep.

Another large meteorite came to Earth in 1908 in Tunguska, Siberia. It did not hit the ground. It exploded in the air. The explosion started fires and blew down trees.

Astronomers also believe that a meteorite may have been the reason the dinosaurs became extinct, or died out.

They believe all the dinosaurs died when an asteroid crashed into Earth about 65 million years ago.

Not all meteorites that hit Earth cause a lot of damage. More than two thousand meteorites have been found. They look like regular rocks. Some are tiny pebbles, and some are large boulders. They are made of stone and iron.

Many meteorites cause little damage and can be studied by astronomers.

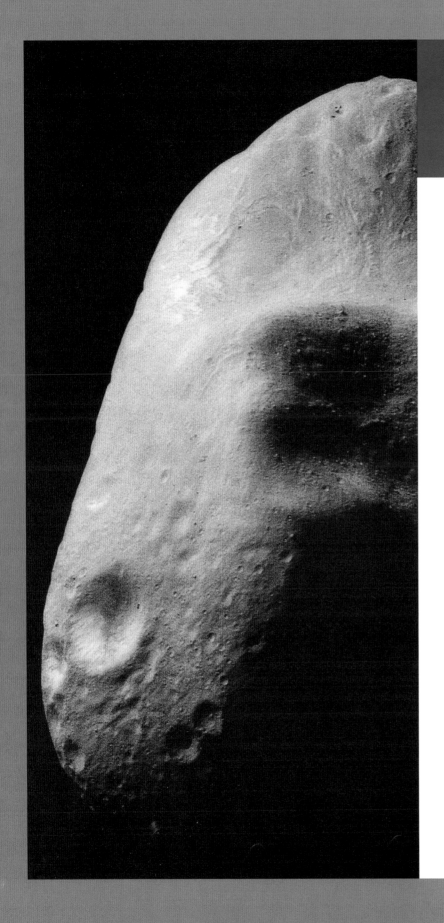

Studying the Solar System

✦ Astronomers have found ways to study these objects floating in space. They watch comets and asteroids closely with their **telescopes**. Spacecraft have been sent to study them as well.

The Near-Earth Asteroid Rendezvous (NEAR) was the first mission to orbit an asteroid. The NEAR *Shoemaker* spacecraft studied the asteroid Eros in 2000 and 2001. Eros is the largest Near-Earth asteroid. The *Shoemaker* orbited the asteroid and took

◀ *The asteroid Eros*

many pictures. Then it landed on the asteroid's surface.

Stardust is on its way to Comet Wild 2. It will pass the comet in 2004 and collect samples of dust and other materials from its coma. Then it will return the samples to Earth in 2006.

Meteorites give astronomers an easy way to study the solar system because they do not have to leave Earth. If a meteorite is from a comet or an asteroid, they can learn a lot about those objects. Meteorites are sometimes pieces of other planets.

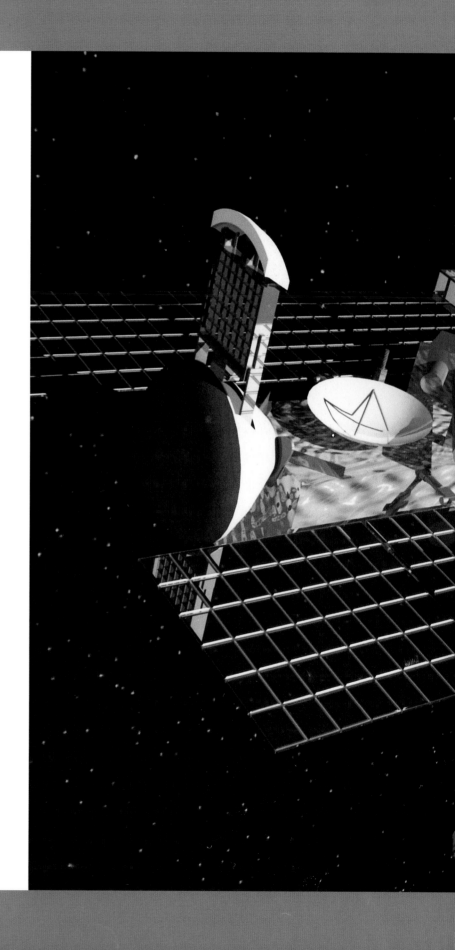

Stardust *will pass Comet* ▶
Wild 2 in 2004 and return
to Earth in 2006.

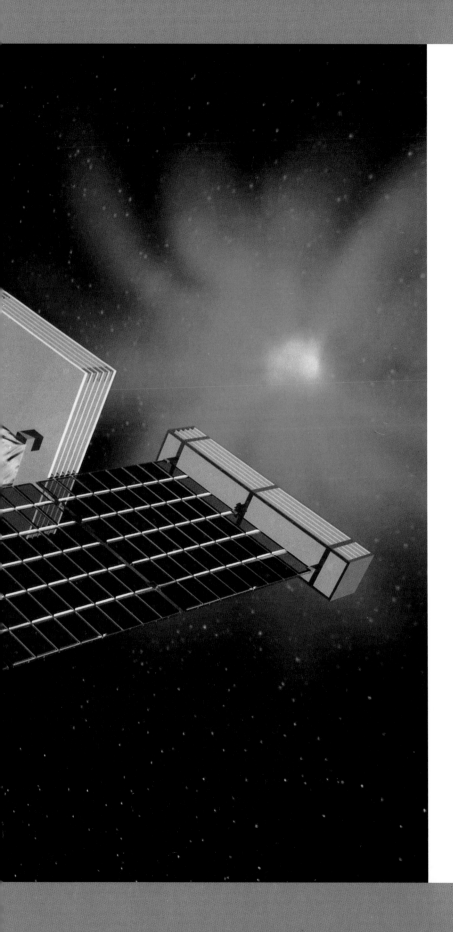

Astronomers believe that meteorites found in Antarctica are from Mars. Other meteorites have been discovered that were once part of the Moon.

Comets, asteroids, and meteoroids are the keys to finding out about our solar system. They are pieces of history. They are bits of the solar system that have been around since it was formed more than 4 billion years ago.

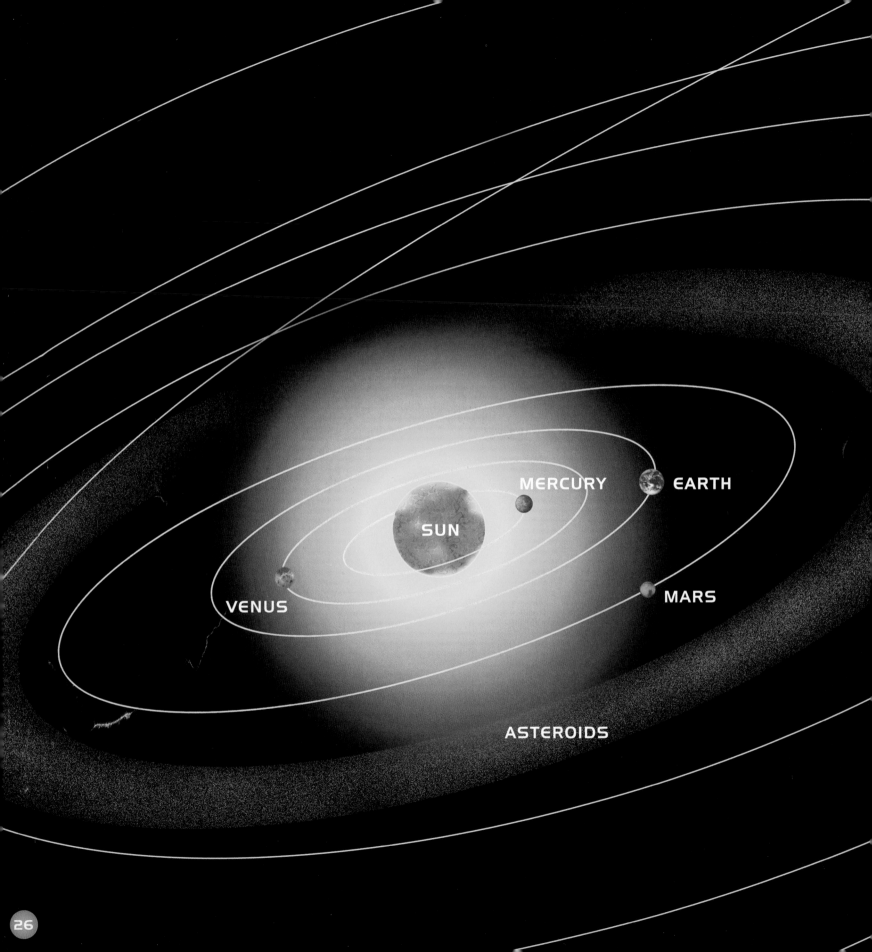

JUPITER

URANUS

SATURN

NEPTUNE

PLUTO

Glossary

asteroids—chunks of rock that orbit the Sun especially between the orbits of Mars and Jupiter

astronomers—people who study space

comets—pieces of ice and rock that have long tails of dust and orbit the Sun

craters—bowl-shaped landforms created by meteorites and asteroids crashing into a planet

gravitational pull—the force inside a planet that pulls objects toward its center

meteoroids—chunks of rock in space; when they hit a planet, they are called meteorites

meteors—pieces of rock that burn up in Earth's atmosphere

telescopes—tools astronomers use to make objects look closer

Did You Know?

- The Kuiper Belt was named after famous astronomer Gerard Kuiper (1905–1973). In 1951, he had the idea that there was an area of comets beyond Pluto. The first object in the Kuiper Belt was not discovered until 1992.

- The Oort Cloud was named after the Dutch astronomer Jan Oort (1900–1992).

- Halley's Comet was last seen in the sky in 1986. It will not be seen from Earth again until 2061.

- Comet Encke has a very short orbit. It takes Encke less than four years to orbit the Sun.

- A comet is pictured in a famous piece of art called the Bayeux Tapestry. The artwork dates from 1066. It is on display in Bayeux, France.

- In December 1994, an NEA came close to Earth. It was within 64,200 miles (103,300 km). That is closer than the Moon to Earth.

- A meteor shower called the Perseids can be seen in the sky every August.

- Meteorites fall toward Earth's surface at about 311 miles (500 km) per hour.

- Meteorites have crashed through people's cars and fallen on their houses. None has ever done major damage.

Want to Know More?

AT THE LIBRARY

Gallant, Roy A. *Comets, Asteroids, and Meteorites.* Tarrytown, N.Y.: Benchmark Books, 2000.

Mitton, Jacqueline and Simon Mitton. *Scholastic Encyclopedia of Space.* New York: Scholastic Reference, 1998.

Redfern, Martin. *The Kingfisher Young People's Book of Space.* New York: Kingfisher, 1998.

Simon, Seymour. *Comets, Meteors, and Asteroids.* New York: Morrow Books, 1994.

ON THE WEB

Near Earth Asteroid Rendezvous
http://nssdc.gsfc.nasa.gov/planetary/near.html
For information and photos about the mission to the asteroid Eros

The Nine Planets
http://www.seds.org/nineplanets/nineplanets/
For a tour of the solar system, including planets, comets, asteroids, and meteoroids

Solar System Exploration
http://sse.jpl.nasa.gov/features/planets/planetsfeat.html
For more information about all the objects in the solar system

Space Kids
http://spacekids.hq.nasa.gov
NASA's space science site designed just for kids

Space.com
http://www.space.com
For the latest news about everything to do with space

Star Date Online
http://stardate.org/resources/ssguide/
For an overview of the solar system

Stardust Mission
http://stardust.jpl.nasa.gov/
For information about this mission to collect material from a comet

THROUGH THE MAIL

Goddard Space Flight Center
Code 130, Public Affairs Office
Greenbelt, MD 20771
To learn more about space exploration

Jet Propulsion Laboratory
4800 Oak Grove Drive
Pasadena, CA 91109
To learn more about spacecraft missions

Lunar and Planetary Institute
3600 Bay Area Boulevard
Houston, TX 77058
To learn more about the planets

Space Science Division
NASA Ames Research Center
Moffet Field, CA 94035
To learn more about solar
system exploration

ON THE ROAD

**Adler Planetarium and
Astronomy Museum**
1300 S. Lake Shore Drive
Chicago, IL 60605-2403
312/922-STAR
To visit the oldest planetarium in
the Western Hemisphere

***Exploring the Planets* and
*Where Next, Columbus?***
National Air and Space Museum
7th and Independence Avenue, S.W.
Washington, DC 20560
202/357-2700
To learn more about the solar system and
space exploration

**Rose Center for Earth and
Space/Hayden Planetarium**
Central Park West at 79th Street
New York, NY 10024-5192
212/769-5100
To visit this new planetarium and learn
more about the solar system

UCO/Lick Observatory
University of California
Santa Cruz, CA 95064
408/274-5061
To see the telescope that was used to
discover the first planets outside of our
solar system

Index

◄ **About the Author:** *Dana Meachen Rau loves to study space. Her office walls are covered with pictures of planets, astronauts, and spacecraft. She also likes to look up at the sky with her telescope and write poems about what she sees. Ms. Rau is the author of more than seventy-five books for children, including nonfiction, biographies, storybooks, and early readers. She lives in Burlington, Connecticut, with her husband, Chris, and children, Charlie and Allison.*